DK eyewonder

Human Body

LONDON, NEW YORK, MUNICH,
MELBOURNE, and DELHI

Written and edited by Caroline Stamps
Designed by Helen Melville
Managing editor Sue Leonard
Managing art editor Cathy Chesson
Category publisher Mary Ling
Picture researcher Marie Osborn
Production Shivani Pandey
DTP designer Almudena Díaz
Consultant Daniel Carter

REVISED EDITION
DK UK
Senior editor Caroline Stamps
Senior art editor Rachael Grady
Jacket design Natasha Rees
Producer (print production) Rita Sinha
Producers (pre-production)
Francesca Wardell, Rachel Ng
Publisher Andrew Macintyre

DK INDIA
Senior editor Priyanka Nath
Senior art editor Rajnish Kashyap
Assistant editor Suneha Dutta
Art editor Isha Nagar
Managing editor Alka Thakur Hazarika
Managing art editor Romi Chakraborty
DTP designer Anita Yadav
Picture researcher Sumedha Chopra

First published in Great Britain in 2003
This edition published in Great Britain in 2013
by Dorling Kindersley Limited
80 Strand, London WC2R 0RL

Copyright © 2003, © 2013 Dorling Kindersley Limited
A Penguin Company

13 14 15 16 17 10 9 8 7 6 5 4 3 2 1
001 – 192754 – 06/13

A CIP catalogue record for this book
is available from the British Library.
ISBN 978-1-4093-2842-1

Colour reproduction by Scanhouse, Malaysia
Printed and bound in China by Hung Hing

Discover more at
www.dk.com

Contents

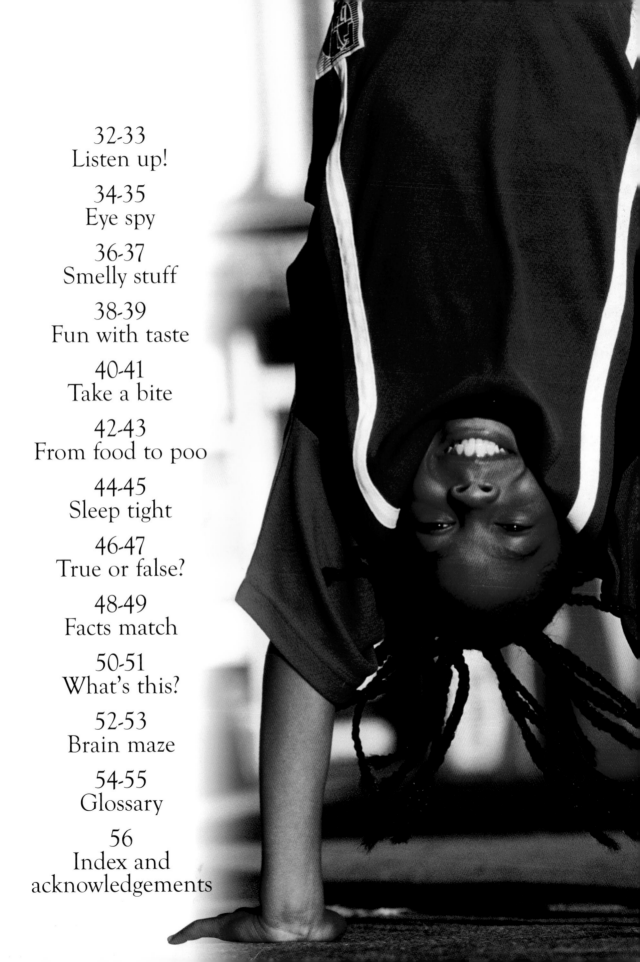

Everyone looks different...

Tall, short, plump, thin, blond, dark... Even though we have two eyes, a nose, two arms, and so on, we still all look so different that we can recognize each person we know without getting anyone confused.

Human beings are different in all sorts of ways. It is thought

Body facts

- The average human body contains enough iron to make a nail 2.5 cm (1 in) long.

- Brown or black skin has more of a pigment called melanin in it than white skin.

- You inherit certain features (such as hair colour or body shape) from your parents.

What about twins?

Only identical twins look alike and that is because they develop at the same time, from one egg that has split into two. Identical twins are always the same sex.

that more than 6,500 languages are spoken throughout the world.

There are slight differences between the left- and right-hand sides of your face.

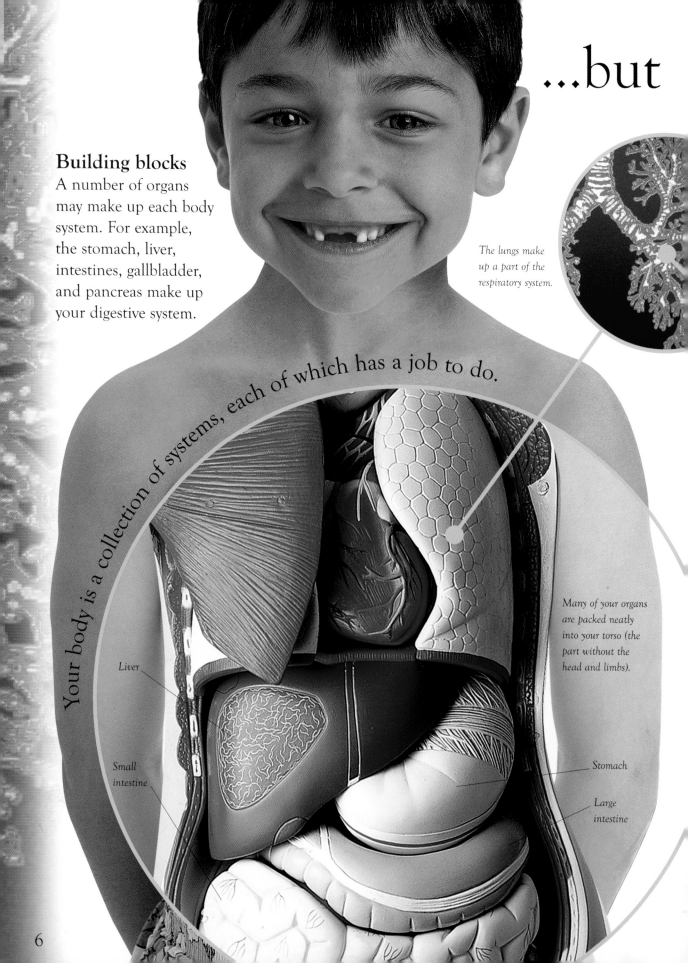

...but

Building blocks

A number of organs may make up each body system. For example, the stomach, liver, intestines, gallbladder, and pancreas make up your digestive system.

The lungs make up a part of the respiratory system.

Your body is a collection of systems, each of which has a job to do.

Many of your organs are packed neatly into your torso (the part without the head and limbs).

Liver

Small intestine

Stomach

Large intestine

we are all alike inside

All bodies are made up of organs. Skin is an organ. It is wrapped around a framework of bones and other organs such as the heart, the brain, and the lungs.

What does an organ do?

Organs work to keep you alive, and each does a different job. Organs work together to make up systems, such as the muscular system and the circulatory system.

It would take about 200 of your cells to fit on a full-stop.

Your body has about 50,000 billion cells.

Made of tissue

Organs are made up of tissue, which is made of groups of similar cells. These magnified cells are from the lungs.

Nucleus

Cell

Different cells

Cells are different depending on the organ they are a part of – skin cells, for example, are different to bone cells. Most cells have a nucleus – the control centre.

Babies and bellybuttons

We all begin life inside our mother as a tiny egg. This develops after it is joined, or fertilized, by a sperm from the father. Most babies spend about 40 weeks growing in their mother's tummy.

Baby facts

● At just eight weeks, the foetus can be recognized as human – although it is shorter than your little finger.

● Fingernails begin to form when the foetus is about ten weeks old.

● A foetus can get hiccups.

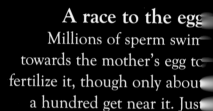

A race to the egg

Millions of sperm swim towards the mother's egg to fertilize it, though only about a hundred get near it. Just one sperm fertilizes it

Legs here, arms there...

After the egg has been fertilized, it begins to divide, becoming a ball of cells. It is full of instructions for what the baby will look like.

It can hear you!

A baby can hear noises from around its mother's tummy – it can hear you talking or laughing, and it will recognize your voice

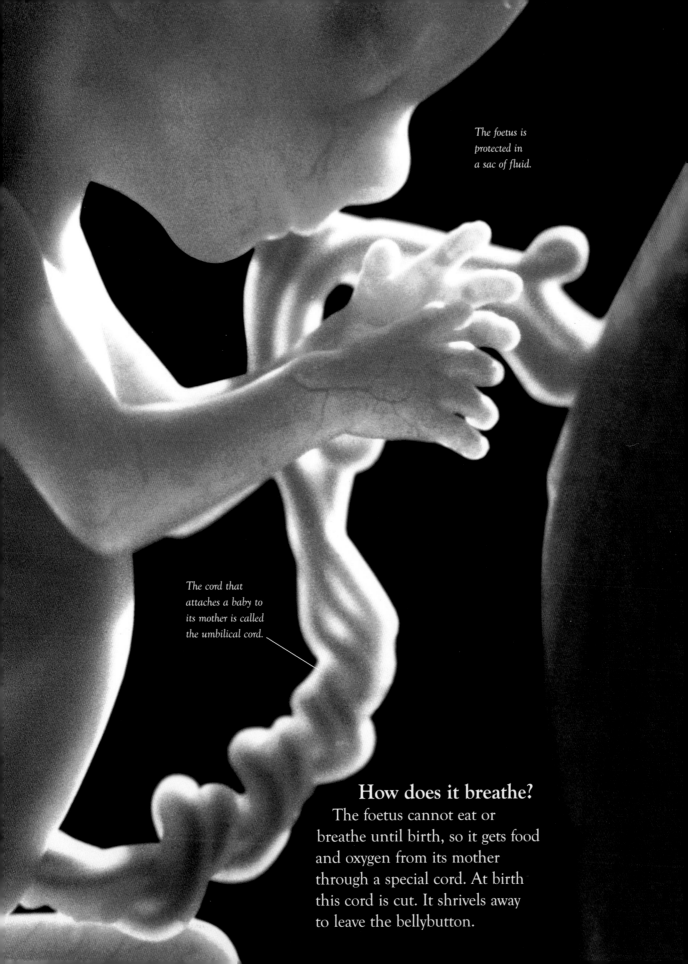

*The foetus is
protected in
a sac of fluid.*

*The cord that
attaches a baby to
its mother is called
the umbilical cord.*

How does it breathe?
The foetus cannot eat or
breathe until birth, so it gets food
and oxygen from its mother
through a special cord. At birth
this cord is cut. It shrivels away
to leave the bellybutton.

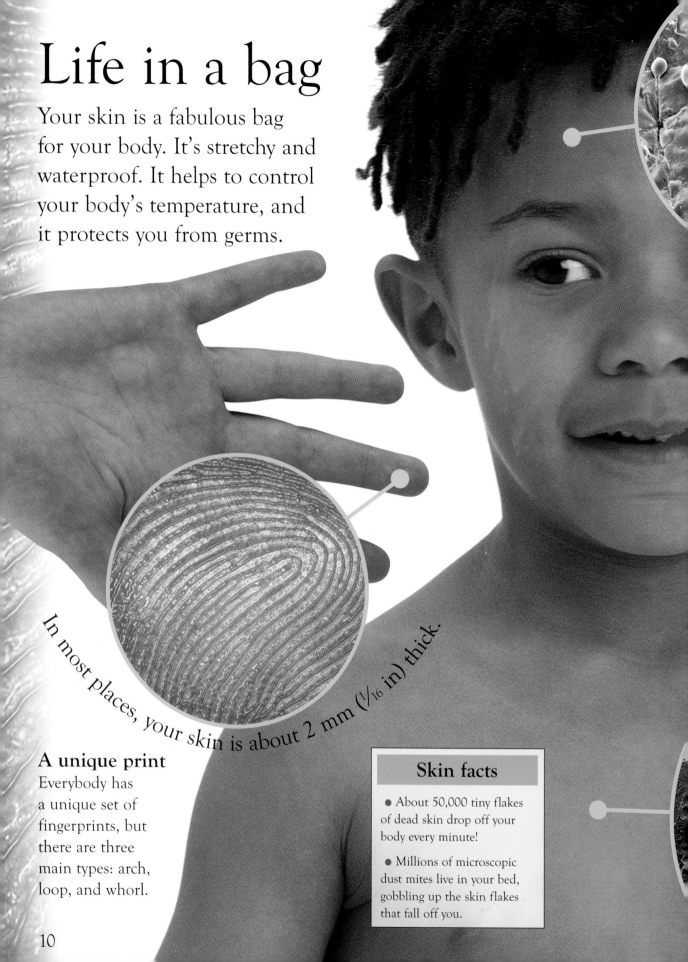

Life in a bag

Your skin is a fabulous bag for your body. It's stretchy and waterproof. It helps to control your body's temperature, and it protects you from germs.

In most places, your skin is about 2 mm ($\frac{1}{16}$ in) thick.

A unique print
Everybody has a unique set of fingerprints, but there are three main types: arch, loop, and whorl.

Skin facts

● About 50,000 tiny flakes of dead skin drop off your body every minute!

● Millions of microscopic dust mites live in your bed, gobbling up the skin flakes that fall off you.

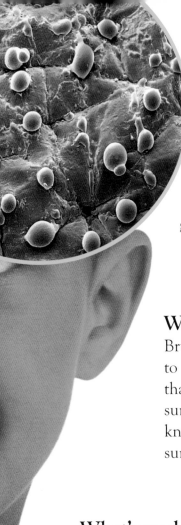

Sweat it off

You sweat to keep cool – but did you know that in a fingernail-sized patch of skin there are between 100 and 600 sweat glands?

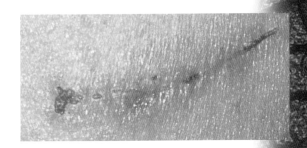

Skin alert...mend that cut!

Cut yourself and a lot of activity in the surrounding skin causes the blood to clot. The resulting scab stops dirt and germs from getting in.

What's a bruise?

Bruises are caused by damage to the tiny blood capillaries that run just under the skin's surface. If broken by a heavy knock, they bleed into the surrounding area.

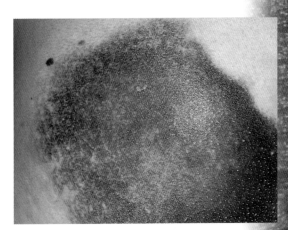

What's underneath?

Skin contains sweat glands, hair follicles, nerve endings, and tiny blood vessels called capillaries. Underneath, there's a layer of fat.

Cells lock together to provide a waterproof layer.

Flakes of dead skin

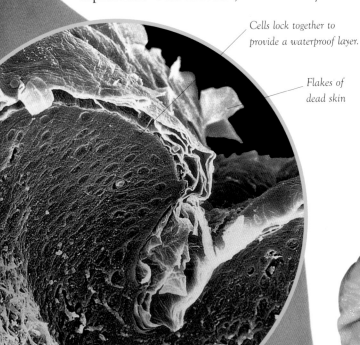

My feet are wrinkly!

Spend a long time swimming and the thicker skin on your feet and hands will begin to wrinkle because water has soaked into it. The extra water makes it pucker up.

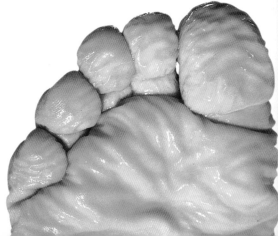

11

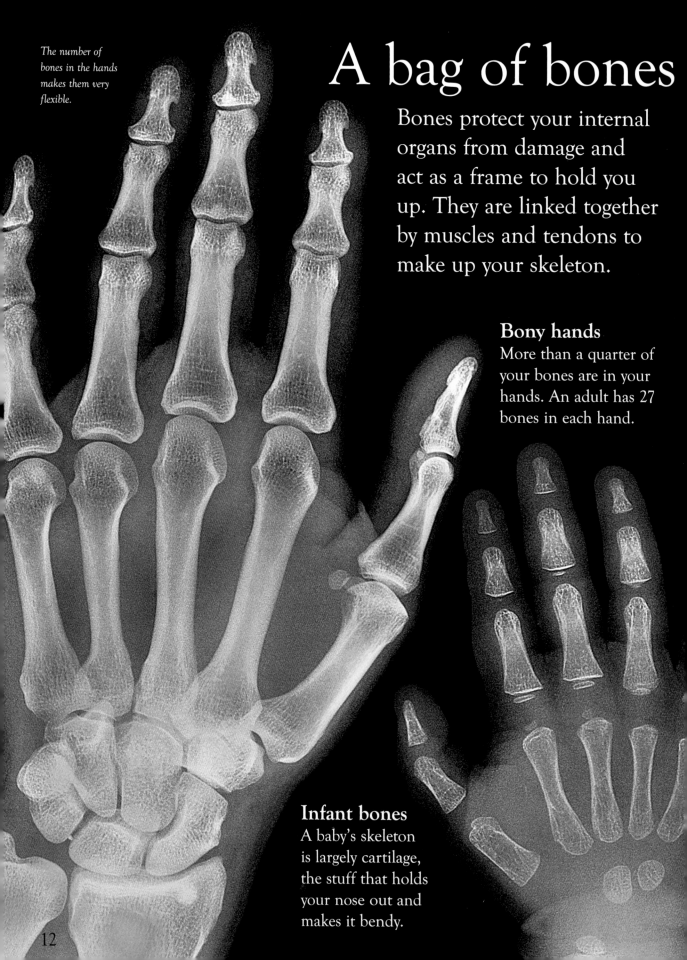

The number of bones in the hands makes them very flexible.

A bag of bones

Bones protect your internal organs from damage and act as a frame to hold you up. They are linked together by muscles and tendons to make up your skeleton.

Bony hands
More than a quarter of your bones are in your hands. An adult has 27 bones in each hand.

Infant bones
A baby's skeleton is largely cartilage, the stuff that holds your nose out and makes it bendy.

It's broken!

If you break a bone, an X-ray shows the doctor what is going on beneath the skin. Bones are living tissue, and will usually mend, with rest and support, in about 6-8 weeks.

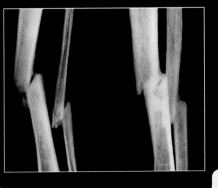

This X-ray shows two broken legs.

An adult skeleton contains 206 bones.

Joints

A joint is the place where two bones meet. This is a hip joint (above), which is a ball-and-socket joint. It helps in movement.

The rounded end of the femur fits snugly into the pelvis.

Pelvis

Femur

Your bones are full of blood vessels, nerves, and cells.

Bone facts

- Compared to a steel bar of the same weight, a bone is far stronger.

- You have the same number of neck bones as a giraffe.

- Bones need calcium from foods like milk and cheese to make them hard.

Hidden support

If you cut through a femur, or thigh bone, you'd see that the inside is a spongy honeycomb. This makes it strong, but light.

13

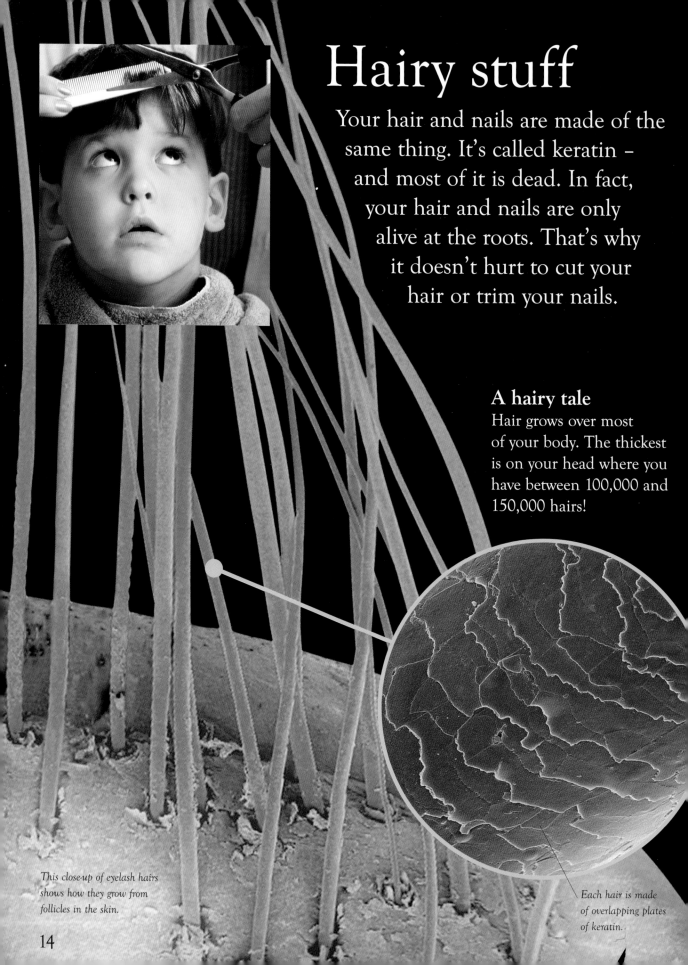

Hairy stuff

Your hair and nails are made of the same thing. It's called keratin – and most of it is dead. In fact, your hair and nails are only alive at the roots. That's why it doesn't hurt to cut your hair or trim your nails.

A hairy tale

Hair grows over most of your body. The thickest is on your head where you have between 100,000 and 150,000 hairs!

This close-up of eyelash hairs shows how they grow from follicles in the skin.

Each hair is made of overlapping plates of keratin.

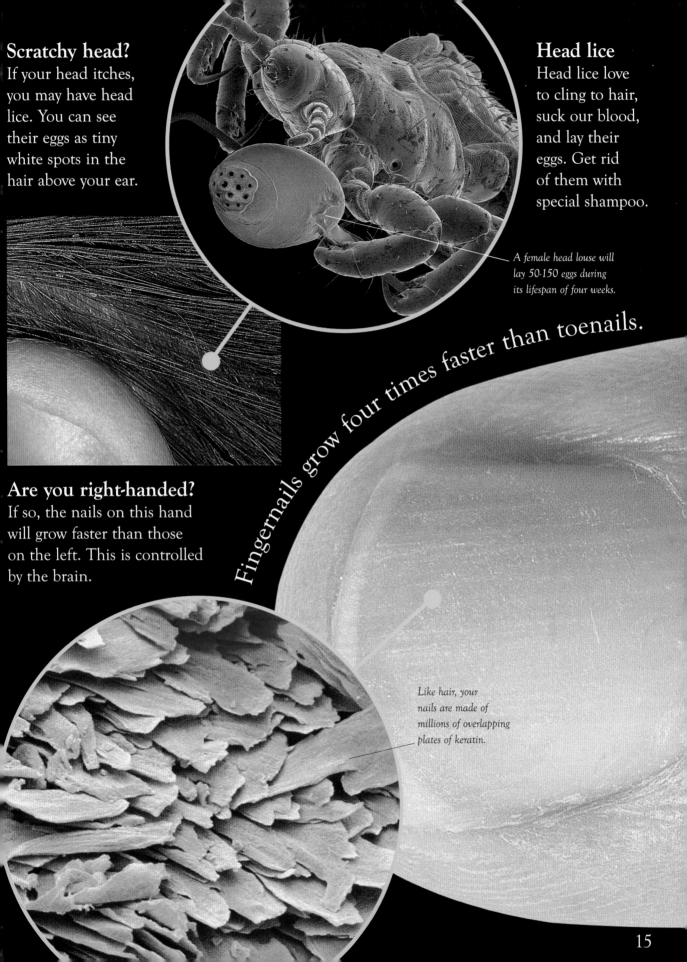

Scratchy head?

If your head itches, you may have head lice. You can see their eggs as tiny white spots in the hair above your ear.

Head lice

Head lice love to cling to hair, suck our blood, and lay their eggs. Get rid of them with special shampoo.

A female head louse will lay 50-150 eggs during its lifespan of four weeks.

Fingernails grow four times faster than toenails.

Are you right-handed?

If so, the nails on this hand will grow faster than those on the left. This is controlled by the brain.

Like hair, your nails are made of millions of overlapping plates of keratin.

Move that body

Step forwards and you'll use about 200 muscles. You have at least 600 muscles, and they are responsible for every movement you make, from jumping to blinking and breathing.

A closer look

A muscle is made up of bundles of tiny fibres. Each fibre is incredibly thin – much thinner than a hair.

Bend your arm

Try tensing the muscle in your upper arm, your biceps. Can you feel it getting harder?

The biceps has contracted.

The triceps has relaxed.

MUSCLE MOUSE

The word "muscle" comes from the Ancient Romans, who thought that muscle movements under the skin looked just like a mouse running about. Their word for mouse was *musculus*.

All joined up

Many muscles are joined to the ends of the bones they control by stringy cords called tendons.

Clench your fist and you can see a tendon working under the skin of your wrist.

Muscle facts

● Your muscles make up 40 per cent of your body's weight.

● Help your muscles grow big and strong by eating plenty of protein. That means lots of eggs, meat, cheese, and beans.

● Muscles can contract to one-third of their size.

Pull a face!

Your face is full of muscles. Incredibly, you use 17 of these muscles to smile. However, you use about 40 muscles to frown!

Biceps and triceps

Muscles can only pull, so they work in pairs. In your arm, the biceps pulls by contracting to bend the arm and the triceps pulls to straighten it.

As a muscle contracts, it gets shorter and harder.
As a muscle relaxes, it gets longer and softer.

The muscles in our face allow us to make about 10,000 different facial expressions!

Pump that blood!

Can you feel your heart beat? This amazing muscle never gets tired, even though it opens and closes about 100,000 times a day, every day, throughout your life.

A one-way system
Your heart beats to push blood around your body. Four valves ensure that the blood always goes the same way.

Where is it?
Your heart is protected by your rib cage. It is slightly to the left of your chest.

A heart has four chambers.

Speed up!
Run and your heart beats faster. This gets more oxygen to your muscles.

Held with string
Heart strings are tiny cords that stop the valves from turning inside out when they close.

18

What is blood?

Blood is made up of a watery liquid called plasma, red cells, white cells, and fragments of cells called platelets.

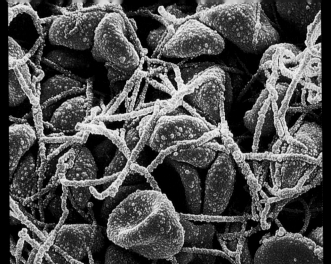

This is what happens when your blood clots because of a cut. The red cells are caught in a mesh of fibres. They die and stop blood flowing out.

The mesh forms very rapidly.

Heart/blood facts

● At rest, a child's heart beats about 85 times a minute.

● A drop of blood contains approximately 250 million red cells, 275,000 white cells, and 16 million platelets.

● A blood cell goes around your body and back through your heart more than 1,000 times each day.

Plasma makes up about 55 per cent of your blood.

Fighting infection

White blood cells and platelets make up less than one per cent of blood. They fight germs.

Red cells are doughnut shaped.

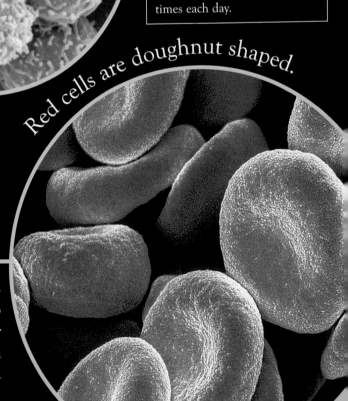

Red blood cells

Red blood cells make up about 44 per cent of your blood. They transport oxygen to various parts of the body.

A circular tale

Your heart pumps blood around your body through arteries and veins. Arteries car ry blood away from the heart. Veins carry blood towards the heart.

Your brain is the hottest part of your body.

Most veins (shown in blue) carry blood that contains carbon dioxide, a waste gas.

Most to the brain
Your brain needs a constant supply of oxygen-rich blood. It is so important that it gets 20 per cent of your body's blood supply.

Blood supply to the brain

Smaller and smaller
Arteries and veins become a branching network of capillaries. The capillary walls are so thin that gases, nutrients, and waste products pass easily through.

Brain

Lung

Lung

Liver

Stomach

Kidney

Kidney

An adult's blood vessels stretch 160,000 km (99,400 miles).

Most arteries (shown in red) carry blood rich with oxygen and food.

Change of colour
As blood travels through the lungs it picks up oxygen. This makes it brighter in colour. As it releases oxygen around the body, it grows darker.

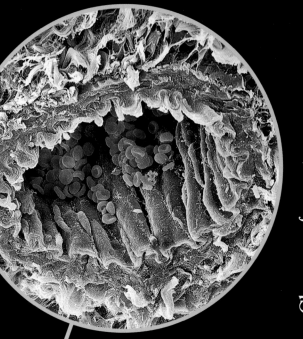

A blood cell travels around your body in about 60 seconds.

Close-up of an artery

This cross-section of an artery is magnified so much that the red blood cells can be seen. Arteries usually have thicker walls than veins.

Your fingers and toes are the coolest parts of your body.

Feel the beat

You can feel your heart's beat as it sends a pulse through the artery in your wrist. Hold your index finger against the inside of your wrist. The regular beat is the surge of pressure that occurs when the heart contracts.

Oxygen-rich blood

Oxygen-poor blood

ONE-WAY SYSTEM

Almost 400 years ago, an English doctor called William Harvey discovered that blood circulates one way around the body, pumped by the heart. Harvey drew detailed diagrams of arteries and veins to show what he meant and published his results in 1628.

Puff, puff

Believe it or not, you take about 23,000 breaths each day. With every breath you take in oxygen, which you need to stay alive, and you breathe out a gas called carbon dioxide, which your body doesn't need.

Windpipe, or trachea

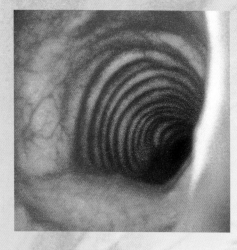

A wind tunnel

Air travels down your windpipe, or trachea, to get to your lungs. In this photograph you can see the rings of cartilage that hold the trachea open.

Taking out the oxygen

The air tubes (shown red) get smaller and smaller until they end in millions of tiny air sacs called alveoli. Here, oxygen is taken into your blood.

Air tube

These spaces are air sacs called alveoli.

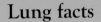

Lung facts

● Stethoscopes, which doctors use to check breathing, were invented in 1816.

● You breathe faster during and after exercise to draw more oxygen into your body.

● Your left lung is smaller than your right lung to allow room for your heart.

Blowing bubbles

We can only store oxygen for a short time in our lungs. Also, unlike fish, we have no gills to remove oxygen from water. So we cannot stay underwater without an air supply.

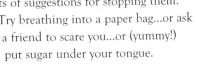

WHY DO I GET HICCUPS?

Hiccups happen when the muscle that helps to move air in and out of your lungs, your diaphragm, jerks uncontrollably. Nobody really knows why they happen, but there are lots of suggestions for stopping them. Try breathing into a paper bag...or ask a friend to scare you...or (yummy!) put sugar under your tongue.

There's water, too

Your breath contains water. If you breathe onto a cold surface, this water condenses into tiny droplets. That means it changes from a vapour into a liquid. The same thing happens on a cold day.

23

Attack of the bugs

Everywhere you go, you are surrounded by nasty germs, and many of them want to live inside your body. After all, it makes a comfy home. The problem is, they can make you ill.

What are germs?
Germs fall into two main groups: bacteria and viruses. Your body is good at keeping them out, but they are clever at finding ways in.

Beastly bacteria
Bacteria come in lots of funny shapes. Some even have tails! If a cut becomes infected (it will look red and swollen), that's be cause bacteria have got in.

Bacteria can double their numbers in 20 minutes!

Vile viruses
Have you had chicken pox? It's caused by a virus. So is the common cold. Viruses are tiny – far smaller than bacteria.

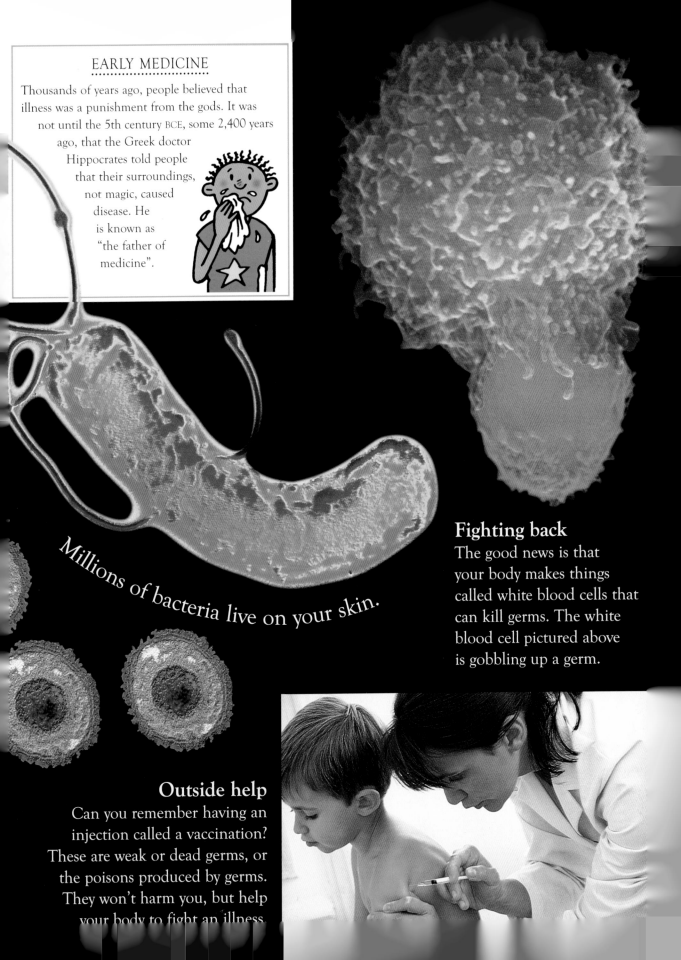

Millions of bacteria live on your skin.

Fighting back

The good news is that your body makes things called white blood cells that can kill germs. The white blood cell pictured above is gobbling up a germ.

Outside help

Can you remember having an injection called a vaccination? These are weak or dead germs, or the poisons produced by germs. They won't harm you, but help your body to fight an illness.

hide

cuddle

giggle

Let's talk

There are many ways of "talking", and not all of them are with your lips. The look on your face and the way you stand tell people a lot about what you are thinking.

I need it now!

Babies can't talk, so they cry to let you know that they want something. From early on, they also communicate by eye contact and facial expression.

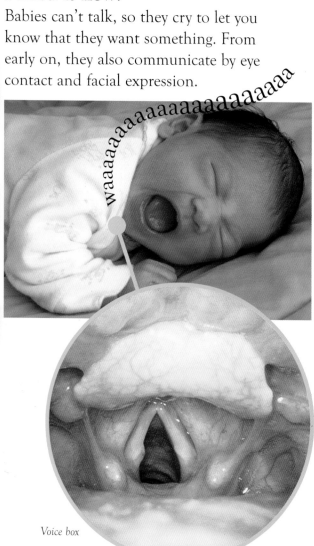

waaaaaaaaaaaaaaaaaaaaaaa

Voice box

Making a word

You make sounds as you breathe out over your voice box, or larynx. Your tongue, lips, and teeth change the sounds into words.

What do you think?

Body language can say a lot about the way you feel. Throw your arms in the air and people know you're excited. Are these children sad?

It is thought that at least 80 per cent of communication is through body language.

Sign language

Signing is one way by which people who are deaf or mute can communicate. They use their hands to sign words and to spell letters.

Some signed words use one hand, others use two.

shout

whisper

cry

Brainbox

Step forward, touch something, talk, drink a glass of milk...everything you do is controlled by your brain. It's a bit like a computer, but far more complicated – and it only weighs 1.3 kg (2.9 lb)!

Sight

Smell

Taste

Touch

Use those senses!
A simple drink requires a lot of brain power. Your eyes and fingers send messages about what you see and touch, while your nose and tongue help you to smell and taste the contents.

Nerves in this girl's fingers "tell" her muscles to grip the glass.

Brain facts

● The brain needs oxygen to work properly. In fact, one-fifth of all the oxygen you breathe in goes to the brain.

● The brain is 85 per cent water.

● The spinal cord stops growing when you are about five years old, having reached about 43 cm (17 in).

How does it work?
Your brain contains billions of nerve cells called neurons that carry signals to and from different parts of your body through your central nervous system.

Why's it so ugly?

Your brain triples in weight between birth and adulthood. As it grows, it wrinkles up to fit your skull, which acts like a protective crash helmet. If you could stretch it out, your brain would cover an ironing board.

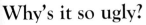

Imagine touching your brain. It would feel like a blancmange!

It's split into two

Your brain has two halves, called hemispheres. The "dominant" half (the one in charge) is usually the left. This is where speech, writing, numbers, and problem-solving are usually handled.

A good fit

The top of a human skull is domed to make room for the brain, as shown by this model.

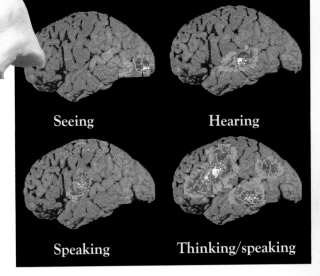

Seeing

Hearing

Speaking

Thinking/speaking

Which bit does what?

Different parts of your brain do different things. These heat scans show which bit of the brain is working for which activity.

29

smooth

slimy

Touch

When you touch something, tiny touch sensors in your skin send a message to your brain.

soft

wet and cold

30

Links to the brain

Sensory nerves carry signals from your skin to your spinal cord, then to your brain. It's your body's branching information system.

Impulses race along some nerves at speeds faster than a racing car.

Nerves are bundles of linked nerve cells. Some nerve cells can be 1 m (3.3 ft) long.

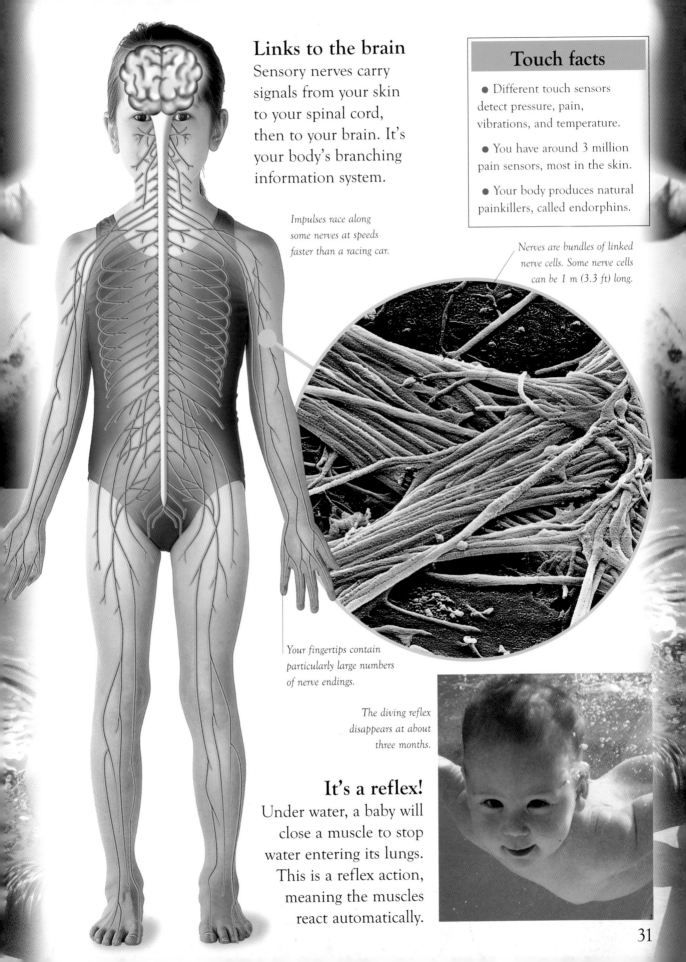

Your fingertips contain particularly large numbers of nerve endings.

The diving reflex disappears at about three months.

It's a reflex!

Under water, a baby will close a muscle to stop water entering its lungs. This is a reflex action, meaning the muscles react automatically.

31

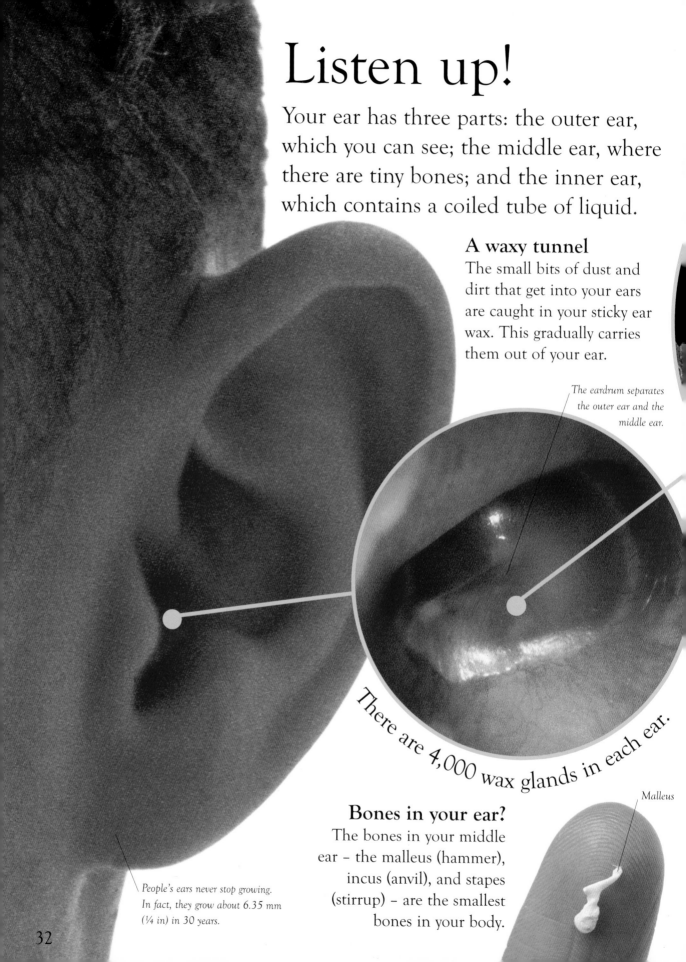

Listen up!

Your ear has three parts: the outer ear, which you can see; the middle ear, where there are tiny bones; and the inner ear, which contains a coiled tube of liquid.

A waxy tunnel

The small bits of dust and dirt that get into your ears are caught in your sticky ear wax. This gradually carries them out of your ear.

The eardrum separates the outer ear and the middle ear.

There are 4,000 wax glands in each ear.

Malleus

Bones in your ear?

The bones in your middle ear – the malleus (hammer), incus (anvil), and stapes (stirrup) – are the smallest bones in your body.

People's ears never stop growing. In fact, they grow about 6.35 mm (¼ in) in 30 years.

Signals travel to the brain along here.

Tiny hairs are moved by sounds.

These tiny hairs are found in the inner ear, in the cochlea. They link up to the brain.

Hairs in your ear?

Tiny hairs in your inner ear pick up movements in the liquid around them. These are sent, as signals, to your brain to "hear".

Ear facts

● Human beings can tell the difference between more than 1,500 different tones of sound.

● Everybody's ears are shaped differently.

● The stapes is the smallest bone in your body; it's shorter than a grain of rice.

Why do I get dizzy?

Your ears tell your brain the position of your head. When you spin, your brain finds it difficult to keep up with the messages sent from your ears. So you feel dizzy.

A little help

If someone is deaf, it means that they cannot hear. A hearing aid helps partially deaf people to hear by making sounds louder.

Eye spy

Those soft, squidgy balls in your head – your eyes – are well protected. They nestle in bony eye sockets and can hide behind your eyelids. Through them your brain receives much of its information about the world.

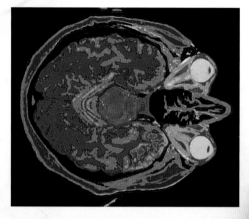

Take a peek inside
This picture shows the two eyes (yellow) in their eye sockets – separated by the nose. They connect directly to the brain.

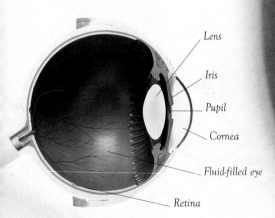

Lens

Iris

Pupil

Cornea

Fluid-filled eye

Retina

What's your colour?
Blue, green, grey, or brown... what colour are your eyes? The colour of your iris depends on the instructions for eye colour that you inherit from your parents.

Your eyes constantly water to keep them free of germs and dust.

A liquid camera
Your eyes are a bit like tiny video cameras, but filled with fluid. Light enters the eye through a hole in the iris, the pupil, and travels to the retina. Messages are sent to the brain, which tells you what you see.

The pupil is smaller in bright light.

The pupil is larger (to let in more light) in dim light.

Your eye can spy a lighted candle 1.6 km (1 mile) away!

How big are your pupils?

Pupil size changes depending on the light – and on what's around you. Do you like what you see? Your pupils will often get bigger. Bored? Your pupils will get smaller.

Eye facts

- You blink about 9,400 times a day.

- Six muscles hold each eye. They are kept busy, moving about 100,000 times a day!

- Microscopic eight-legged mites live in the base of your eyelashes. A yucky fact? Not really – they eat up nasty germs for you.

What is colour blindness?

Your retina contains pigments that detect colour. If these are not working, you will have difficulty telling some colours apart. This is known as colour blindness.

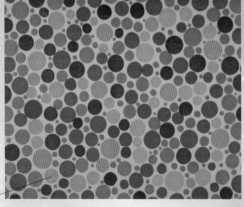

Can you see this number? If not, the pigment that picks up red light may be missing from your retina.

Smelly stuff

Did you know that humans have the ability to tell the difference between about 10,000 smells? This incredible sense helps you to taste and enjoy things.

Cilia

How do we smell?

Things have a smell because they give off particles called molecules. Sniff something and these travel up to cilia at the top of your nose. Under the microscope, cilia look like tiny hairs.

Smell receptors

When the molecules reach the top of your nose, they dissolve in the mucus (or snot) that your nose constantly produces. They then travel to the smell receptors.

Cells at the top of your nose produce about 1 litre (2 pints) of mucus a day.

You have 10 million smell receptors.

A path to the brain

The smell receptors then send a message to your brain, which either recognizes the smell or remembers it if it hasn't come across that smell before.

A smell is recognized in an area towards the front of the brain.

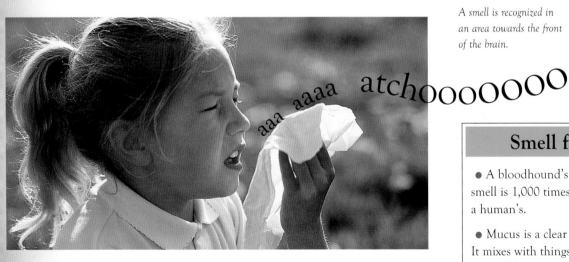

aaa aaaa atchooooOOOO

Why do flowers make me sneeze?

If you have an allergic reaction to pollen, too much mucus will pour into your nose to try and flush it out. There's so much that you have to sneeze to get rid of it.

Smell facts

● A bloodhound's sense of smell is 1,000 times better than a human's.

● Mucus is a clear fluid. It mixes with things in the air and they give it a colour.

● The mucus in your nose can become green if you have an infection.

Fun with taste

Have you ever wondered what your tongue does? It helps you to talk, but it also helps you to move food around your mouth and, more importantly, to taste it.

Have a sniff

Smell plays an important part when you taste a food. That's why things don't taste so good if you have a blocked nose.

If your frenulum is short, you will not be able to stick your tongue out very far.

Ten thousand taste buds help you to tell the difference between five different flavours.

Anchored in place

A flap of skin called the frenulum holds the bottom of your tongue to the floor of your mouth. It stops you from swallowing your tongue.

Tastes on your tongue

When food enters your mouth, bits dissolve in saliva. The different flavours that the tongue can detect are sweet, sour, bitter, salty, and umami (a pleasant savoury taste associated with, among others, cheese and tomatoes).

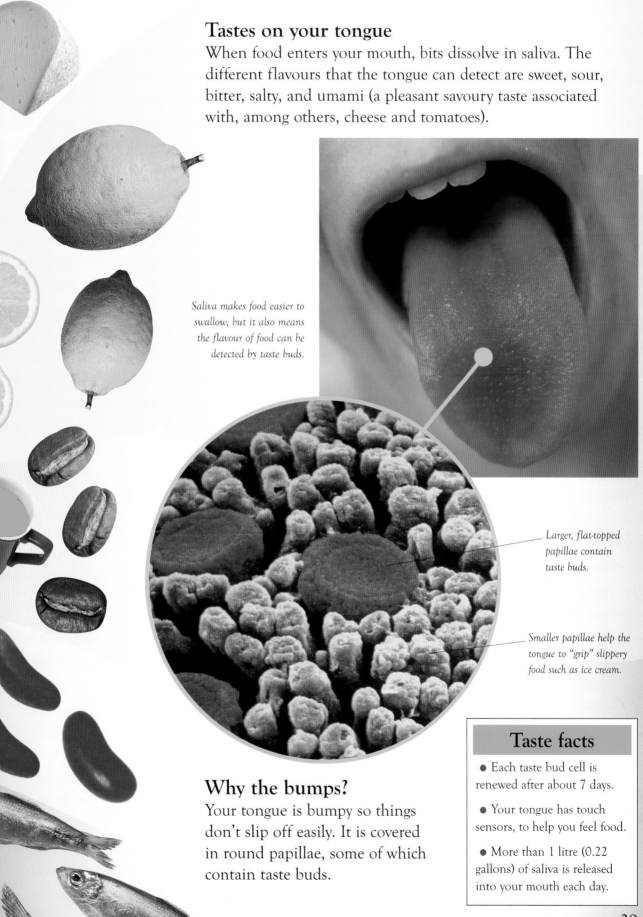

Saliva makes food easier to swallow, but it also means the flavour of food can be detected by taste buds.

Larger, flat-topped papillae contain taste buds.

Smaller papillae help the tongue to "grip" slippery food such as ice cream.

Why the bumps?

Your tongue is bumpy so things don't slip off easily. It is covered in round papillae, some of which contain taste buds.

Taste facts

● Each taste bud cell is renewed after about 7 days.

● Your tongue has touch sensors, to help you feel food.

● More than 1 litre (0.22 gallons) of saliva is released into your mouth each day.

Take a bite

Before their first teeth appear, babies drink milk or eat puréed food. Without teeth, they cannot chew on food to make it easier to swallow. Teeth are very important.

The large knobbly teeth at the back are molars.

Incisor

Molar

A child has 20 milk teeth.

There are 32 adult teeth.

How big are they?
Each of your teeth has a long root, which holds it tightly in your jaw. Inside each tooth are nerves and blood vessels.

Canines are slightly pointed. They help to tear food.

Teeth are rooted in your gums.

Why do they fall out?

Your milk teeth are your first teeth, but they can't grow. So they are pushed out between the ages of 6 and 12 to make room for your adult teeth.

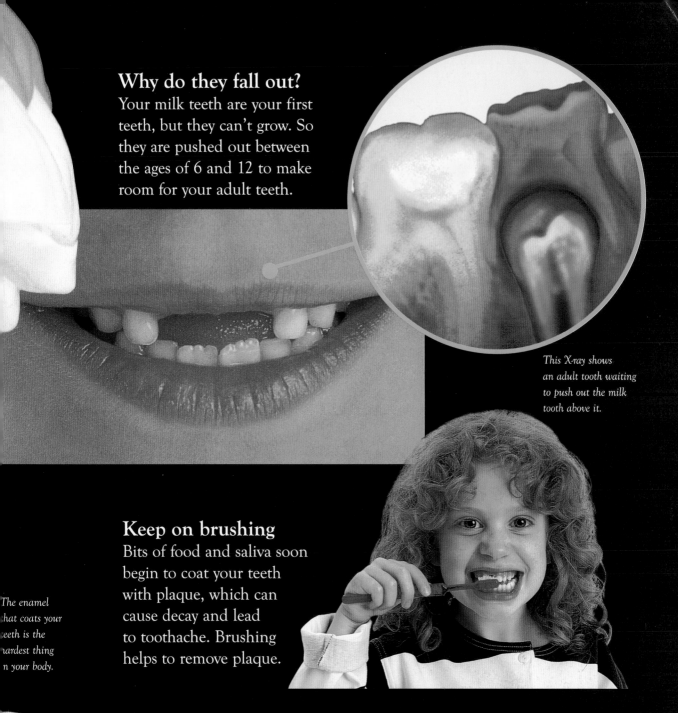

This X-ray shows an adult tooth waiting to push out the milk tooth above it.

Keep on brushing

Bits of food and saliva soon begin to coat your teeth with plaque, which can cause decay and lead to toothache. Brushing helps to remove plaque.

The enamel that coats your teeth is the hardest thing in your body.

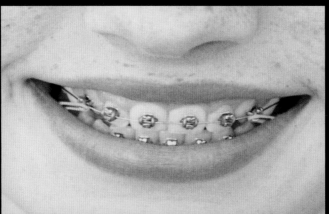

Why do I need braces?

Sometimes your teeth grow crookedly. Braces help to straighten them, making them sit evenly in your mouth.

The braces put a gentle pressure on each tooth.

From food to poo

Food gives us many things, including the energy to run and jump. Energy is also used to break down or digest the food we eat. The nutrients this releases are passed to our cells through the bloodstream. Cells use nutrients to make more energy.

An acid bath

Acid is released in your stomach to break down the food. A constant churning helps turn the food into a mushy soup.

Food travels from your mouth to your stomach in the time it takes you to read this sentence – about ten seconds.

Going down

After you have chewed your food, it is pushed down a tube called the oesophagus and into your stomach.

The sphincter muscle lets food out of your stomach.

You munch your way through some 500 kg (1,100 lb) of food each year. That's the weight of a small car.

Taking the nutrients

The small intestine is lined with finger-like villi. Blood runs through these where it can pick up goodies from the food and take them to the liver. The liver keeps what your body needs.

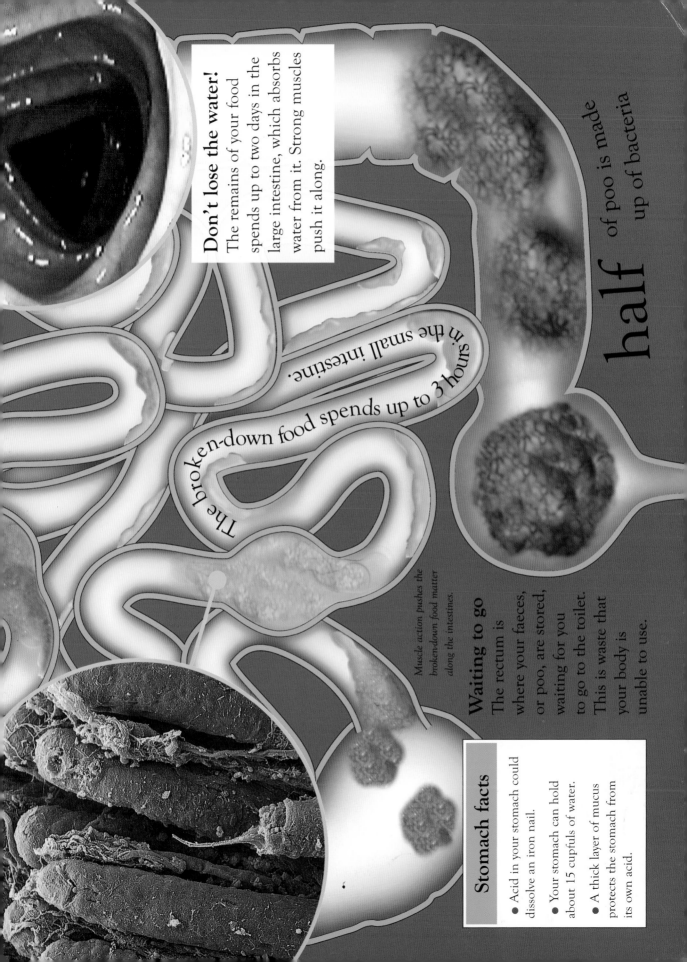

Don't lose the water!

The remains of your food spends up to two days in the large intestine, which absorbs water from it. Strong muscles push it along.

The broken-down food spends up to 3 hours in the small intestine.

half of poo is made up of bacteria

Muscle action pushes the broken-down food matter along the intestines.

Waiting to go

The rectum is where your faeces, or poo, are stored, waiting for you to go to the toilet. This is waste that your body is unable to use.

Stomach facts

- Acid in your stomach could dissolve an iron nail.
- Your stomach can hold about 15 cupfuls of water.
- A thick layer of mucus protects the stomach from its own acid.

Sleep tight

After all the activities you do each day, your body needs to rest. Sleep gives your brain a chance to catch up with what you've done. Without it you cannot think properly and your body will begin to slow down.

Why do I yawn?

If you are bored or sleepy your breathing slows. You yawn to pull more oxygen into your body, helping to keep you awake.

Miss a night's sleep and you'll be cross and clumsy the next day.

A five-year-old needs about ten hours of sleep a night.

You wriggle about a lot when you're sleeping, changing position about 45 times a night.

WHERE ARE YOU GOING?

Sometimes people walk in their sleep. They may even get dressed, or try to find something to eat. But when they wake up in the morning, they won't remember anything about it. More children sleepwalk than adults, and more boys than girls. Nobody really knows why people sleepwalk, but it is usually harmless.

What was that?

Children sometimes have frightening dreams called nightmares, usually about being chased. Remember, nightmares are not real.

Why do I dream?

Dreams bring pictures of things you have seen during the day, but also images that are unrelated to the day's events. Nobody knows exactly why people dream.

It can be noisy!

Snoring happens if a person cannot move air easily through the nose and mouth during sleep. It causes a loud noise.

ZZZZZZZ

A growth hormone is released when a child is asleep.

zzzzzzzzZZZZZZZZ

45

True or false?

It's time to test your knowledge about your body. Spot whether these statements are true or false. Have a go!

After an egg is fertilized, it begins to divide.
See page 8

A red blood cell takes a day to move around your body.
See page 19

Plasma makes about 90 per cent of your blood.
See page 19

Your breath contains water.
See page 23

The top of a skull is domed to make room for the brain.
See page 29

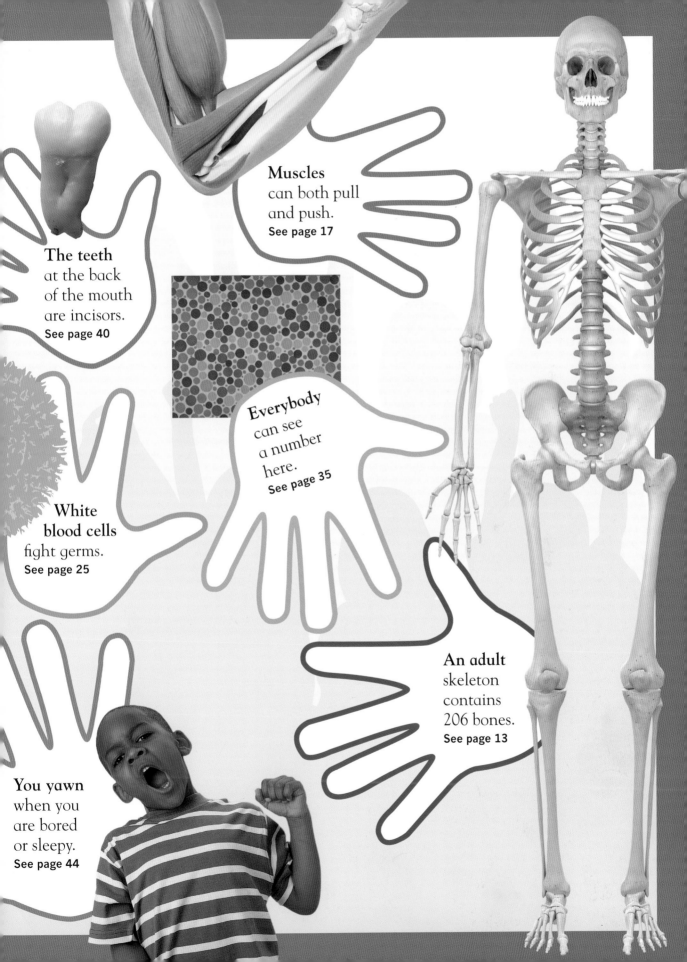

Muscles can both pull and push.
See page 17

The teeth at the back of the mouth are incisors.
See page 40

Everybody can see a number here.
See page 35

White blood cells fight germs.
See page 25

An adult skeleton contains 206 bones.
See page 13

You yawn when you are bored or sleepy.
See page 44

Facts match

What parts of the body are these clues describing? The answers can all be found in the picture.

This is made up of the same stuff as your nails. When you snip it off with scissors, it doesn't hurt. **See page 14**

If you cut through this, you will see that its insides look like a spongy honeycomb. It is strong, yet light at the same time. **See page 13**

There are 600 of these "mice" in your body, which you need for every movement you make. **See page 16**

This organ is a bag for your body. It helps to control your body's temperature, and protects you from germs. **See page 10**

Shaped like doughnuts, these make up 44 per cent of your blood. **See page 19**

Food gets turned into a mushy soup inside here. **See page 42**

This part of your body never stops growing. It can grow about 6.35 mm (¼ in) in 30 years. **See page 32**

There are lots of these places in your body, wherever bones meet. They give you lots of movement. **See page 13**

Hair

Eye

Brain

Ear

Joints

Lungs

Heart

Muscles

Stomach

Small intestine

Nerves

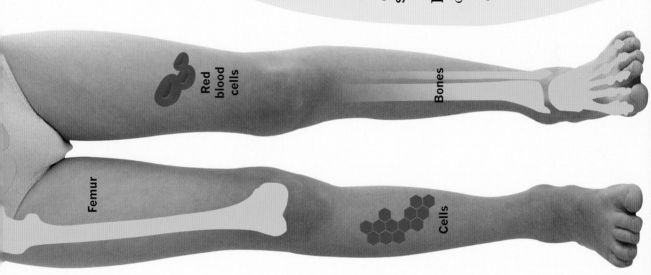

Fingerprints

These need calcium from milk and cheese to stay strong and hard. See page 13

There are 50,000 billion of these in your body. Each of these have a control centre called the nucleus. **See page 7**

Everybody has their own unique set of these patterns. **See page 10**

This is like a tiny video camera filled with fluid. It is directly connected to the brain. **See page 34**

Red blood cells

Femur

Bones

Cells

Skin

Your body has two of these, but the left one is smaller than the right one. **See pages 22–23**

This organ is a bit like a computer for your body and 85 per cent of it is made up of water. **See page 28**

These are your body's messengers, carrying signals from your skin to your spinal cord and brain. **See page 31**

Arteries and veins are dependent on this organ. While veins supply it with blood, arteries carry blood away from it to different parts of the body. **See page 20**

Broken-down food stays in this part of the digestive system for up to 3 hours. **See page 43**

What's this?

Take a look at these close-ups of pictures in the book, and see if you can identify them. The clues should help you!

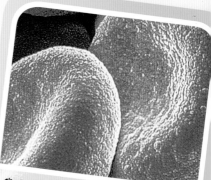

- They are shaped like doughnuts.
- There are about 250 million of these in a drop of blood.

See page 19

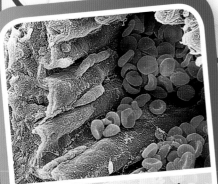

- It takes blood away from the heart.
- It carries blood rich with oxygen and food.

See pages 20–21

- As it grows, it wrinkles to fit your skull.
- It is split into two halves, called hemispheres.

See page 29

- The top part is shaped like a dome.
- It is made like that to make room for the brain.

See page 29

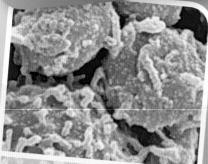

- They fight against germs.
- They make up less than one per cent of your blood.

See page 19

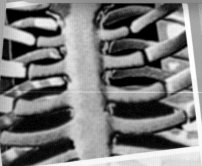

- It contains 206 bones.
- It is made up of bones that are linked together by muscles and tendons.

See pages 12–13

50

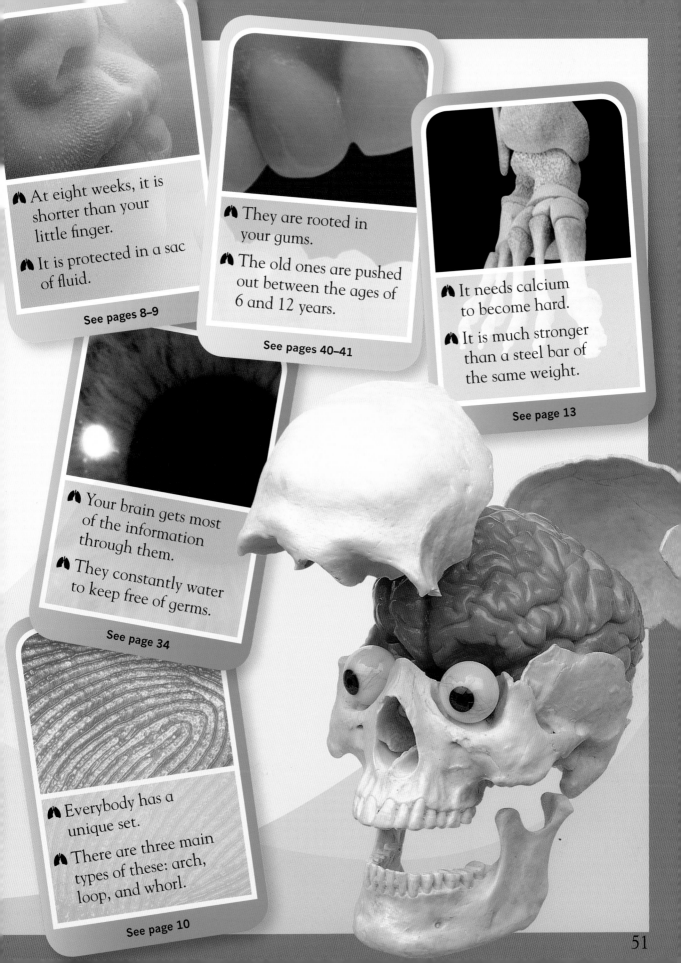

At eight weeks, it is shorter than your little finger.

It is protected in a sac of fluid.

See pages 8–9

They are rooted in your gums.

The old ones are pushed out between the ages of 6 and 12 years.

See pages 40–41

It needs calcium to become hard.

It is much stronger than a steel bar of the same weight.

See page 13

Your brain gets most of the information through them.

They constantly water to keep free of germs.

See page 34

Everybody has a unique set.

There are three main types of these: arch, loop, and whorl.

See page 10

Brain maze

How brainy are you? Find the right answers to the questions below to work your way through the brain maze.

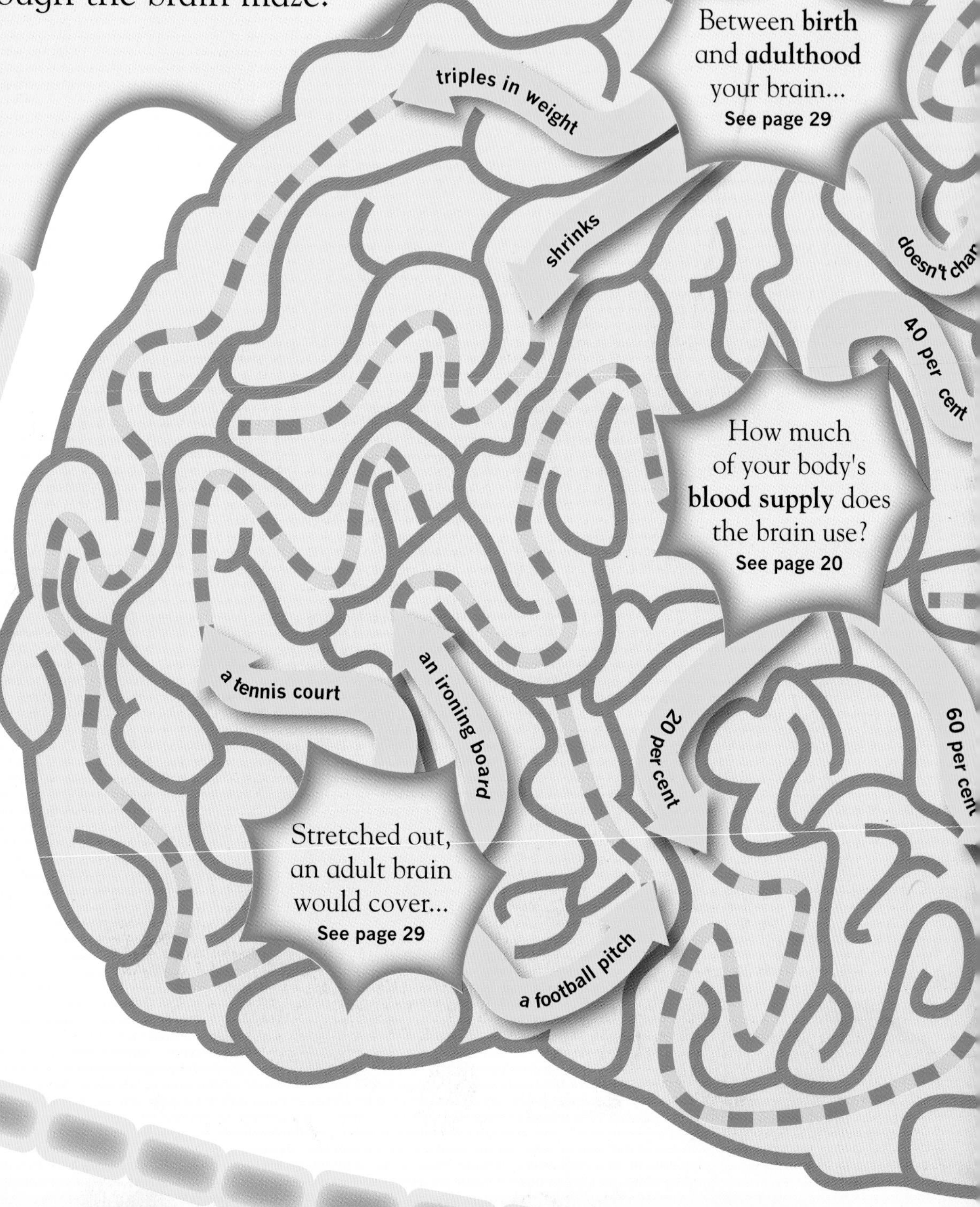

triples in weight

shrinks

Between **birth** and **adulthood** your brain...
See page 29

doesn't chan[ge]

40 per cent

How much of your body's **blood supply** does the brain use?
See page 20

a tennis court

an ironing board

20 per cent

60 per cent

Stretched out, an adult brain would cover...
See page 29

a football pitch

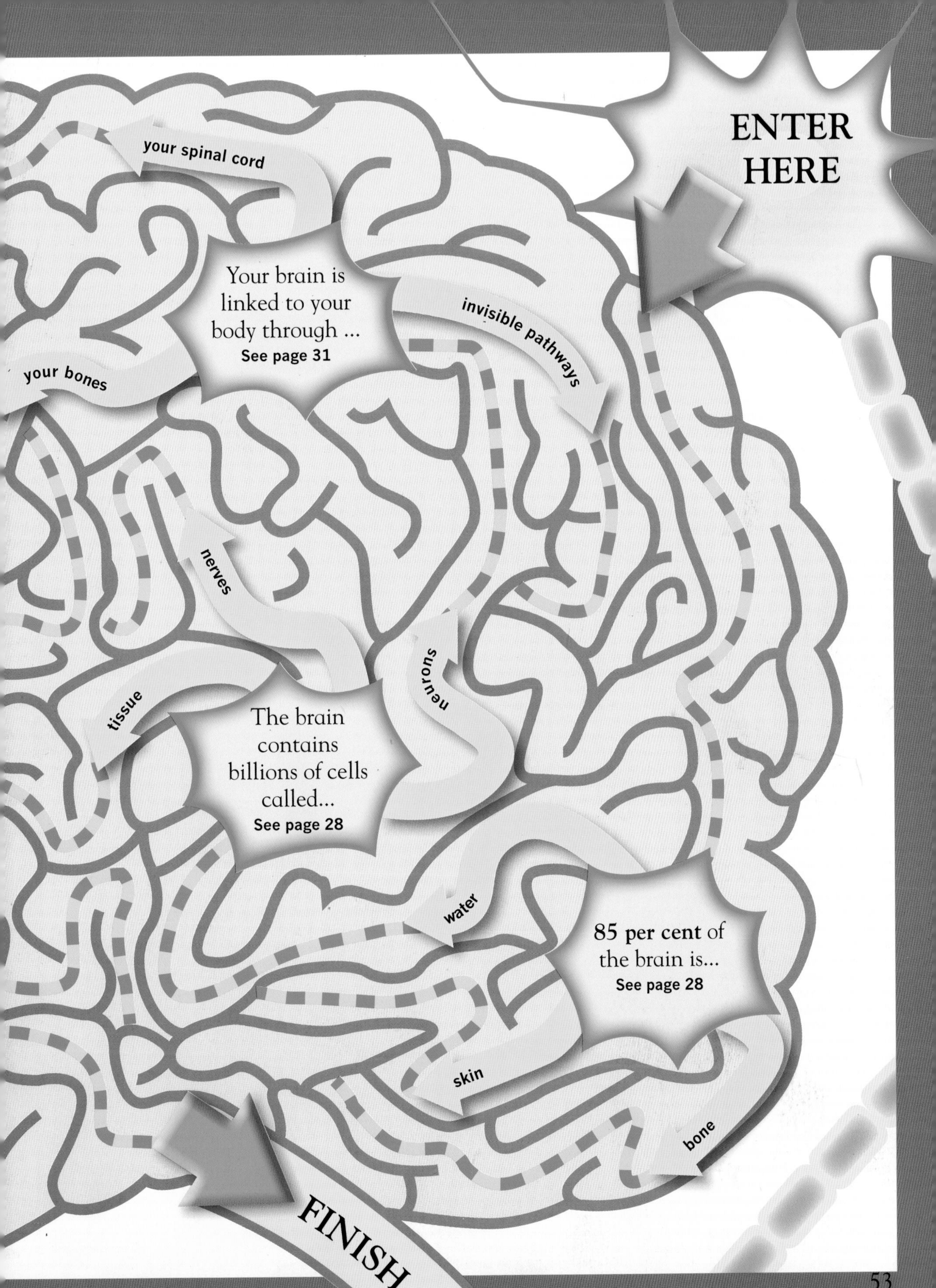

ENTER HERE

your spinal cord

Your brain is linked to your body through ...
See page 31

invisible pathways

your bones

nerves

tissue

neurons

The brain contains billions of cells called...
See page 28

water

85 per cent of the brain is...
See page 28

skin

bone

FINISH

Glossary

Here are the meanings of some words that are useful to know when learning about the human body.

Alveoli microscopic airbags inside the lungs. These are where oxygen from air breathed in is passed into the blood.

Artery part of the network of vessels that carry blood around the body. Arteries carry blood away from the heart.

Blood vessel one of the arteries, veins, and capillaries that carry blood through the body.

Carbon dioxide the waste gas that humans breathe out.

Cartilage tough but flexible material that makes up much of a baby's skeleton. Smaller amounts are found in an adult's body.

Cell one of the body's basic building blocks.

Central nervous system the part of the body's communication system that consists of the brain and the spinal cord.

Diaphragm the muscle that stretches across the chest just below the lungs, helping a person to breathe.

Digestion the process of breaking down food.

Faeces the solid waste that is produced by digestion.

Foetus the unborn baby nine weeks after fertilization.

Germs microscopic bacteria and viruses that cause sickness.

Intestine the long tube through which food passes in the process of digestion.

Larynx the part of the throat where speech sounds are made.

Mucus a slippery fluid that is found in areas such as the respiratory and digestive systems.

Muscle a tissue that contracts to cause movement.

Nerve a bundle of fibres through which instructions pass between different areas and cells in the body.

Nutrients substances in food that are useful to the body (such as proteins, carbohydrates, and vitamins).

Oesophagus the tube that runs between the throat and the stomach.

Organ one of a number of different parts of the body that each perform a particular job.

Oxygen the gas that humans take from air. Oxygen is needed to release energy from food.

Papillae tiny bumps on the surface of the tongue.

Plasma the part of blood that remains when the red and white cells are removed.

Pore tiny holes in the skin through which the body sweats.

Reflex an automatic action, such as breathing or blinking.

Saliva a fluid released into the mouth that helps begin the breakdown of food and makes it slippery enough to swallow.

Senses the means by which humans find out about the world around them. The five senses are hearing, sight, taste, touch, and smell.

Spinal cord the bundle of nerves that runs inside the backbone.

Sweat a liquid that contains waste products. It is released through pores in the skin to help the body cool down.

Tendon a tough cord that links muscle to bone.

Trachea the tube that runs from the larynx to the lungs.

Umbilical cord the cord that connects a foetus to its mother through the placenta.

Vaccination an injection of dead or weak germs, or the toxins produced by germs, that teaches the body to fight that particular germ.

Vein part of the network of vessels that carry blood around the body. Veins carry blood towards the heart.

Vertebra one of the bones that make up the backbone.

Villi Finger-like projections from the wall of the small intestine through which nutrients are taken into the blood.

Voice box see larynx.

Index

Acknowledgements

Dorling Kindersley would like to thank: Penny Arlon for editorial assistance, Dorian Spencer Davies and Andrew O'Brien for original artwork, and Sonia Whillock for design assistance.

Picture credits

The publisher would like to thank the following for their kind permission to reproduce their photographs::
(Key: a-above; b-below/bottom; c-centre; f-far; l-left; r-right; t-top)

Corbis: 14tl, 26cr, 27tr, 31br, 35br, 38l; **Lester V. Bergman** 47c; momentimages/**Tetra Images** 4-5bc **Dorling Kindersley: Donks Models** 58tl; **Jeremy Hunt - modelmaker** 46crb, 51br; **The Natural History Museum, London** 50crb, 59br; **Foodpix:** 28; **Getty Images:** 2-3, 4tr, 5tr, 26bl, 27br, 30bl, 33bl, 36, 44-5, 46-7, 48; **Tom Pfeiffer/VolcanoDiscovery/**

Photographer's Choice 48tr; **ImageState:** 1, 11br, 17cr, 23tr, 27cla, 30tl, 30tr, 30br, 39tr; **Age Fotostock** 4-5b; **Imagingbody.com:** 13cb, 15br; **Masterfile UK:** 41bl; **Allen Birnbach** 26tl; **Robert Karpa** 45tl; **Gail Mooney** 26cl; **Brian Pieters** 27cr; **PunchStock: Brand X Pictures** 51tc; **Science Photo Library:** 6-7t, 7cl, 7br, 8tr, 8ca, 8bl, 8br, 9, 10c, 10-11t, 11tr, 11cr, 11bl, 11r; **St. Stephen's Hospital** 12, 13tc, 13cl, 13r, 14bl, 14br, 15tc, 15cl, 15bl, 16c, 17tl, 18tr, 18cl, 18br, 18, 19tc, 19c, 19br; **Susumu Nishinaga** 21tc, 22tl, 22tc, 22cl, 22bl, 22-23, 24tl, 24bl, 24-25b, 24-25t, 25tr, 25br, 26br, 31cr, 33tl, 34-5, 35tc, 35tr, 36cr, 37tl, 37cr, 37bl, 42tl, 42-43b, 43bl, 46cl, 50tr; **Geoff Tompkinson** 29cl, 50c; **Roger Harris** 47r, 50bc; **SIU** 51cl; **David Becker** 51bl; **Edelmann** 51tl; **Steve Gschmeissner** 50cl, 50bl; **BSIP/VEM** 40tr; **Gregory Dimijian** 34cla; **Pascal Goetgheluck** 40; **Mehau Kulyk** 29tr; **Omikron** 39c; **Wellcome Dept. of Cognitive Neurology** 29br; **The Wellcome Institute Library, London:** 32cr.

All other images © Dorling Kindersley
For further information, see www.dkimages.com

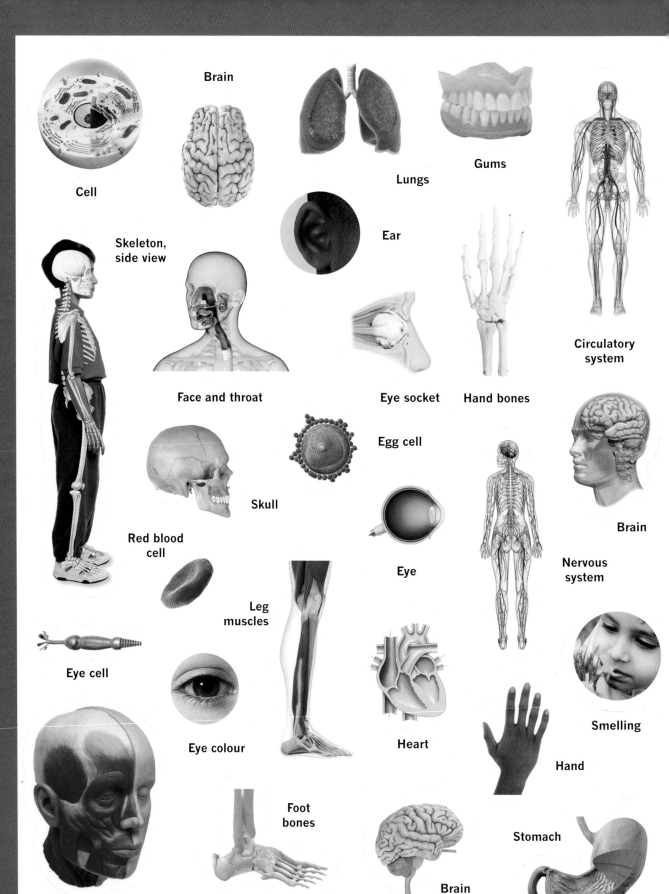

Cell

Brain

Lungs

Gums

Skeleton,
side view

Ear

Circulatory
system

Face and throat

Eye socket

Hand bones

Egg cell

Skull

Brain

Red blood
cell

Eye

Nervous
system

Eye cell

Leg
muscles

Eye colour

Heart

Smelling

Hand

Foot
bones

Stomach

Brain

Face muscles

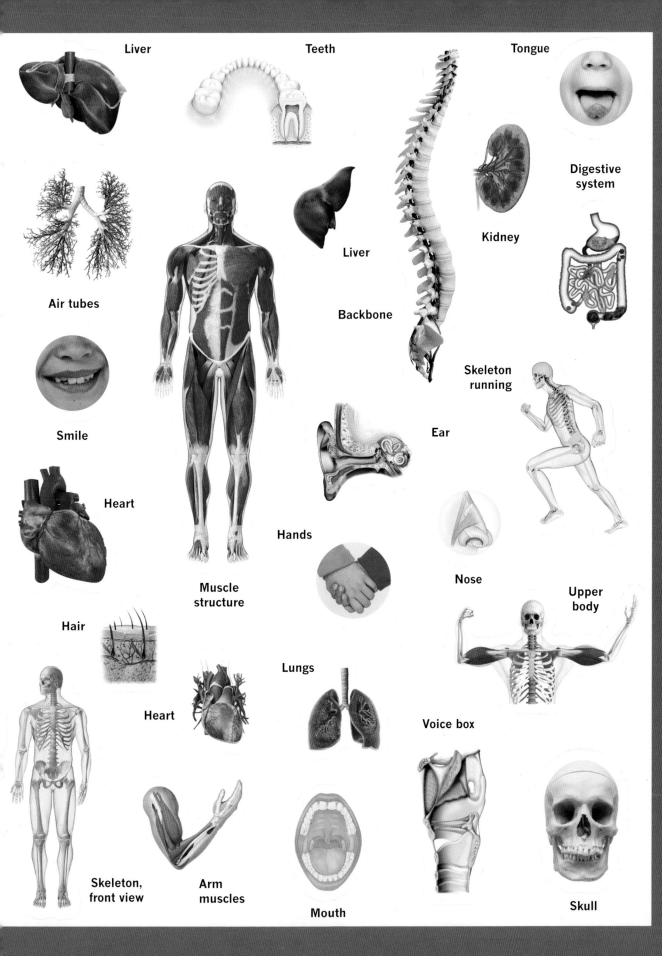

Liver

Teeth

Tongue

Digestive
system

Kidney

Air tubes

Liver

Backbone

Smile

Skeleton
running

Heart

Ear

Hands

Nose

Muscle
structure

Upper
body

Hair

Lungs

Heart

Voice box

Skeleton,
front view

Arm
muscles

Mouth

Skull